1 MONTH OF
FREE
READING

at

www.ForgottenBooks.com

By purchasing this book you are eligible for one month membership to ForgottenBooks.com, giving you unlimited access to our entire collection of over 1,000,000 titles via our web site and mobile apps.

To claim your free month visit:

www.forgottenbooks.com/free1224429

ISBN 978-0-332-70419-7
PIBN 11224429

This book is a reproduction of an important historical work. Forgotten Books uses
state-of-the-art technology to digitally reconstruct the work, preserving the original format
whilst repairing imperfections present in the aged copy. In rare cases, an imperfection in
the original, such as a blemish or missing page, may be replicated in our edition. We do,
however, repair the vast majority of imperfections successfully; any imperfections that
remain are intentionally left to preserve the state of such historical works.

Excerpts from newspapers and other sources

From the files of the

Lincoln Financial Foundation Collection

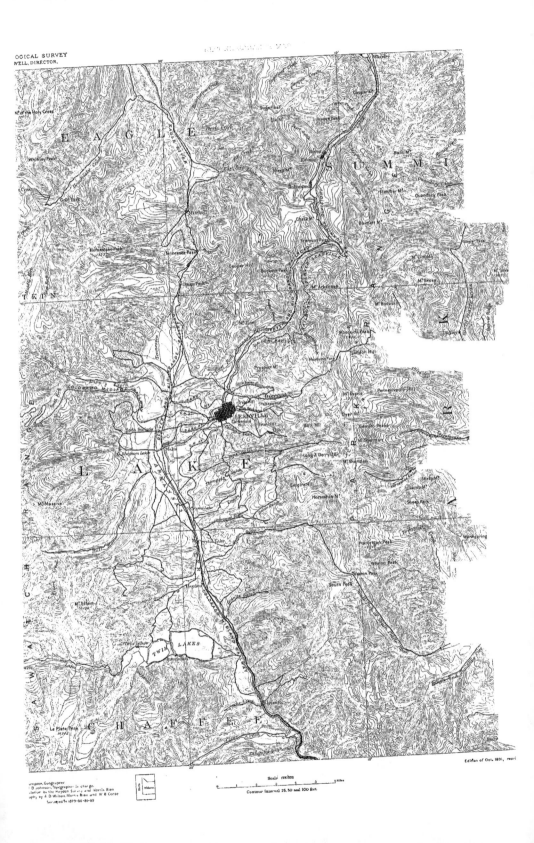

EAGLE

SUMMI

T K I N

LAKE

CHAFFEE

M? of the Holy Cross

Whitney Peak

Homestake Peak

Tennessee Pass

M? Massive

La Plata Peak

Sugarloaf

Robinson

Rhone

Buckeye Peak

M? Zion

LEADVILLE

Bartlett M?

M? Sheridan

Weston Peak

South Peak

Ptarmigan Peak

TWIN LAKES

Wheeler

Bald M?

Fletcher M?

Quandary Peak

M? Lincoln

M? Sherman

unipson, Geographer
D Johnson, Topographer in charge.
lation by the Hayden Survey and Morris Bien
aphy by E. D. Wilson, Morris Bien and W. B. Corse
Surveyed in 1879-80-81-89

Scale 125000
0   1   2   3   4   5   6 Miles
Contour Interval 25, 50 and 100 feet.

True Meridian

DESCRIPTION OF THE TOPOGRAPHIC MAP OF THE UNITED STATES

The United States Geological Survey is making a topographic map of the United States. This work has been in progress since 1882, and about three-tenths of the area of the country, excluding outlying possessions, has been mapped. The mapped areas are widely scattered, nearly every State being represented, as shown on the progress map accompanying each annual report of the Director.

This great map is being published in atlas sheets of convenient size, which are bounded by parallels and meridians. The four-cornered division of land corresponding to an atlas sheet is called a *quadrangle*. The sheets are of approximately the same size: the paper dimensions are 20 by 16½ inches; the map occupies about 17½ inches of height and 11½ to 16 inches of width, the latter varying with latitude. Three scales, however, are employed. The largest scale is 1:62500, or very nearly one mile to one inch; i. e., one linear mile on the ground is represented by one linear inch on the map. This scale is used for the thickly settled or industrially important parts of the country. For the greater part of the country an intermediate scale of 1:125000, or about two miles to one inch, is employed. A third and still smaller scale of 1:250000, or about four miles to one inch, has been used in the desert regions of the far West. A few special maps on larger scales are made of limited areas in mining districts. The sheets on the largest scale cover 15′ of latitude by 15′ of longitude; those on the intermediate scale, 30′ of latitude by 30′ of longitude; and those on the smallest scale, 1° of latitude by 1° of longitude.

The features shown on this map may, for convenience, be classed in three groups: (1) *water*, including seas, lakes, ponds, rivers and other streams, canals, swamps, etc.; (2) *relief*, including mountains, hills, valleys, cliffs, etc.; (3) *culture*, i. e., works of man, such as towns, cities, roads, railroads, boundaries, etc. The conventional signs used for these features are grouped below. Variations appear in some maps of earlier dates.

All water features are shown in *blue*, the smaller streams and canals in full blue lines; and the larger streams, lakes, and the sea by blue water-lining. Certain streams, however, which flow during only a part of the year, their beds being dry at other times, are shown, not by full lines, but by lines of dots and dashes. Ponds which are dry during a part of the year are shown by oblique parallel lines. Salt-water marshes are shown by horizontal ruling interspersed with tufts of blue, and fresh-water marshes and swamps by blue tufts with broken horizontal lines.

Relief is shown by contour lines in *brown*. Each contour passes through points which have the same altitude. One who follows a contour on the ground will go neither uphill nor downhill, but on a level. By the use of contours not only are the shapes of the plains, hills, and mountains shown, but also the elevations. The line of the seacoast itself is a contour line, the datum or zero of elevation being mean sea level. The contour line at, say, 20 feet above sea level is the line that would be the seacoast if the sea were to rise or the land to sink 20 feet. Such a line runs back up the valleys and forward around the points of hills and spurs. On a gentle slope this contour line is far from the present coast line, while on a steep slope it is near it. Thus a succession of these contour lines far apart on the map indicates a gentle slope; if close together, a steep slope; and if the contours run together in one line, as if each were vertically under the one above it, they indicate a cliff. In many parts of the country are depressions or hollows with no outlets. The contours of course surround these, just as they surround hills. Those small hollows known as sinks are usually indicated by hachures, or short dashes, on the inside of the curve. The contour interval, or the vertical distance in feet between one contour and the next, is stated at the bottom of each map. This interval varies according to the character of the area mapped; in a flat country it may be as small as 10 feet; in a mountainous region it may be 200 feet. Certain contours, usually every fifth one, are accompanied by figures stating elevation above sea level. The heights of many definite points, such as road corners, railroad crossings, railroad stations, summits, water surfaces, triangulation stations, and bench marks, are also given. The figures in each case are placed close to the point to which they apply, and express the elevation to the nearest foot only. The exact elevations of bench marks and

their descriptions, as well as the descriptions and geodetic coordinates of triangulation stations, are published in the annual reports and bulletins of the Survey. The publications pertaining to specified localities may be had on application.

The works of man are shown in *black*, in which color all lettering also is printed. Boundaries, such as State, county, city, land-grant, reservation, etc., are shown by broken lines of different kinds and weights. Cities are indicated by black blocks, representing the built-up portions, and country houses by small black squares. Roads are shown by fine double lines (full for the better roads, dotted for the inferior ones), trails by single dotted lines, and railroads by full black lines with cross lines. Other cultural features are represented by conventions which are easily understood.

The sheets composing the topographic atlas are designated by the name of a principal town or of some prominent natural feature within the district, and the names of adjoining published sheets are printed on the margins. The sheets are sold at five cents each when fewer than 100 copies are purchased, but when they are ordered in lots of 100 or more copies, whether of the same sheet or of different sheets, the price is three cents each.

The topographic map is the base on which the facts of geology and the mineral resources of a quadrangle are represented. The topographic and geologic maps of a quadrangle are finally bound together, accompanied by a description of the district, to form a folio of the Geologic Atlas of the United States. The folios are sold at twenty-five cents each, except such as are unusually comprehensive, which are priced accordingly.

Applications for the separate topographic maps or for folios of the Geologic Atlas should be accompanied by the cash or by post-office money order (not postage stamps), and should be addressed to—

THE DIRECTOR,

United States Geological Survey,

Washington, D. C.

July, 1905.

CONVENTIONAL SIGNS

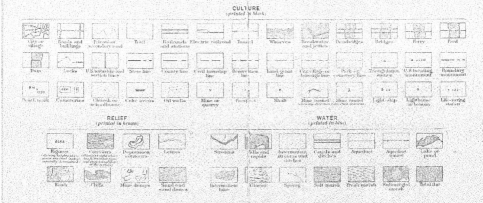

CULTURE
(printed in black)

City or village — Roads and buildings — Private or secondary road — Trail — Railroads and stations — Electric railroad — Tunnel — Wharves — Breakwater and jetties — Drawbridge — Bridges — Ferry — Ford

Dam — Locks — U.S. township and section lines — State line — County line — Civil township line — Reservation line — Land-grant line — City village or borough line — Park or cemetery line — Triangulation station — U.S. bench mark monument — Boundary monument

Bench mark — Cemeteries — Church or schoolhouse — Coke ovens — Oil wells — Mine or quarry — Prospect — Shaft — Mine tunnel — Mine mouth — Light ship — Lighthouse or beacon — Life-saving station

RELIEF
(printed in brown)

Figures — Contours — Depression contours — Levees — Streams — Falls and rapids — Intermittent streams and lake line — Canals and ditches — Aqueduct — Aqueduct tunnel — Lake or pond

Wash — Cliffs — Mine dumps — Sand and sand dunes — Intermittent lake — Glacier — Spring — Salt marsh — Fresh marsh — Submerged marsh — Tidal flat

WATER
(printed in blue)

GEOLOGIC AND OTHER REPORTS

The following reports relate to Colorado and New Mexico but are not parts of the topographic or geologic atlas. An asterisk (*) indicates that the report is out of print, but many such reports are available for consultation in certain libraries. (See list on p. 8.) The publications for which the price is stated are sold by the Superintendent of Documents, Government Printing Office, Washington, D. C. Remittance to that official should be made by postal money order, express order, or check; postage stamps will not be accepted.

ANNUAL REPORTS:

*Second, 1880–81. Contains: Abstract of report on geology and mining industry of Leadville, Colo., by S. F. Emmons. pp. 201–290.

Sixth. 1884–85. Contains: Mount Taylor and the Zuni Plateau, by C. E. Dutton, pp. 105–198. $2.

Eighth, 1886–87. I contains: The fossil butterflies of Florissant, Colo., by S. H. Scudder, pp. 433–474. $1.50.Part

Ninth, 1887–88. Contains: On the geology and physiography of a portion of northwestern Colorado and adjacent parts of Utah and Wyoming, by C. A. White, pp. 677–712. $2.

Thirteenth, 1891–92. Part III contains: Report upon the construction of topographic maps and the selection and survey of reservoir sites in the hydrographic basin of Arkansas River, Colo., by A. H. Thompson, pp. 429–444. $1.85.

Fourteenth, 1892–93. *Part II contains: The laccolitic mountain groups of Colorado, Utah, and Arizona, by Whitman Cross, pp. 157–241.

Sixteenth, 1894–95. Part II contains: Geology and mining industries of the Cripple Creek district, Colo., by Whitman Cross and R. A. F. Penrose, jr., pp. 1–209; Water resources of a portion of the Great Plains. by Robert Hay, pp. 535–585. $1.25.

Seventeenth, 1895–96. *Part II contains: Geology of Silver Cliff and the Rosita Hills, Colo., by Whitman Cross, pp. 263–403; The mines of Custer County, Colo., by S. F. Emmons, pp. 405–472; The underground water of the Arkansas Valley in eastern Colorado, by G. K. Gilbert, pp. 551–601.

Eighteenth. 1896–97. Part I contains: Triangulation and spirit leveling, by H. M. Wilson and others, pp. 131–422; $1. *Part III contains: Preliminary report on the mining industries of the Telluride quadrangle, Colo., by C. W. Purington, pp. 745–850.

Nineteenth, 1897–98. Part I contains: Triangulation and spirit leveling, by H. M. Wilson and others, pp. 145–408. $1.

Twentieth, 1898–99. Part I contains: Triangulation and spirit leveling, by H. M. Wilson and others, pp. 211–530; $1. Part II contains: Devonian fossils from southwestern Colorado; The fauna of the Ouray limestone, by G. H. Girty, pp. 25–81; $2.50. Part V contains: Pikes Peak, Plum Creek, and South Platte Reserves, by J. G. Jack, pp. 39–115; White River Plateau Timber Land Reserve, by G. B. Sudworth, pp. 117–179, and Battlement Mesa Forest Reserve, by G. B. Sudworth, pp. 181–243; $2.80.

Twenty-first, 1899–1900. *Part II contains: Geology of the Rico Mountains, Colo., by Whitman Cross and A. C. Spencer, pp. 7–165. Part IV contains: The High Plains and their utilization, by W. D. Johnson, pp. 601–741; $2.35.

Twenty-second, 1900–1901. *Part I contains: The asphalt and bituminous rock deposits of the United States, by G. H. Eldridge, pp. 209–452. *Part II contains: The ore deposits of the Rico Mountains, Colo., by F. L. Ransome, pp. 229–398. *Part III contains: The Rocky Mountain coal fields, by L. S. Storrs, pp. 415–471. Part IV contains: The High Plains and their utilization, by W. D. Johnson, pp. 631–869; $2.20.

MONOGRAPHS:

*12. Geolgy and mining industry of Leadville, Colo., by S. F. Emmons. 1886. 770 pp.

*27. Geology of the Denver Basin in Colorado, by S. F. Emmons and others. 1896. 556 pp.

31. Geology of the Aspen mining district, Colo., by J. E. Spurr. 1898. 260 pp. $3.60.

35. The later extinct floras of North America, by J. S. Newberry. 1898. 295 pp. $1.25.

40. Adephagous and clavicorn Coleoptera from the Tertiary deposits at Florissant, Colo., by S. H. Scudder. 1900. 148 pp. 80c.

44. Pseudoceratites of the Cretaceous, by Alpheus Hyatt, 1903. 351 pp. $1.

49. The Ceratopsia, by J. B. Hatcher. 1907. 300 pp. $1.50.

51. Cambrian Brachiopoda, by C. D. Walcott. 1912. In two parts. Part I, 872 pp.; Part II, 363 pp. $4.

54. The Mesozoic and Cenozoic Echinodermata of the United States, by W. B. Clark and M. W. Twitchell. 1915. 341 pp. $1.50.

PROFESSIONAL PAPERS:

*16. The Carboniferous formations and faunas of Colorado, by G. H. Girty. 1903. 546 pp.

*32. Preliminary report on the geology and underground water resources of the central Great Plains, by N. H. Darton. 1905. 433 pp.

*33. Forest conditions in the Lincoln Forest Reserve, N. Mex., by F. G. Plummer and M. G. Gowsell. 1904. 47 pp.

*39. Forest conditions in the Gila River Forest Reserve, N. Nex., by T. F. Rixon. 1905. 89 pp.

*52. Geology and underground waters of the Arkansas Valley in eastern Colorado, by N. H. Darton. 1906. 90 pp.

*54. Geology and gold deposits of the Cripple Creek district, Colo., by Waldemar Lindgren and F. L. Ransome. 1906. 516 pp.

*63. Economic geology of the Georgetown quadrangle (together with the Empire district), Colo., by J. E. Spurr and G. H. Garrey. 1908. 422 pp.

67. Landslides in the San Juan Mountains, Colo., by Ernest Howe. 1909. 58 pp. 25c.

*68. The ore deposits of New Mexico, by Waldemar Lindgren and others. 1910. 361 pp.

75. Geology and ore deposits of the Breckenridge district, Colo., by F. L. Ransome 1911 187 pp. 75c.

*90. Shorter contributions to general geology, 1914. Contains *(a) Geology of the pitchblende ores of Colorado. by E. S. Bastin. pp. 1–5; *(c) Dike rocks of the Apishapa quadrangle, Colo., by Whitman Cross, pp. 17–31; (r) Contributions to the stratigraphy of southwestern Colorado, by Whitman Cross and E. S. Larsen, pp. 39–50, 10c.; *(k) The history of a portion of Yampa River, Colo.. and its possible bearing on that of Green River, by E. T. Hancock, pp. 188–189.

*93. Geology of the Navajo country—a reconnaissance of parts of Arizona, New Mexico, and Utah, by H. E. Gregory. 1917. 161 pp.

*94. Economic geology of Gilpin County and adjacent parts of Clear Creek and Boulder Counties, Colo, by E. S. Bastin and J. M. Hill. 1917. 379 pp.

*95. Shorter contributions to general geology, 1915. Contains: (b) Eocene glacial deposits in southwestern Colorado, by W. W. Atwood, pp. 13–26, 15c.; *(c) Relation of the Cretaceous formations to the Rocky Mountains in Colorado and New Mexico, by W. T. Lee, pp. 27–58.

*98. Shorter contributions to general geology, 1916. Contains: *(h) The flora of the Fox Hills sandstone. by F. H. Knowlton. pp. 85–93; (p) Stratigraphy of a part of the Chaco River valley, N. Mex., by C. M. Bauer. pp. 271–278, 15c.; (q) Vertebrate faunas of the Ojo Alamo. Kirtland, and Fruitland formations, by C. W. Gilmore, pp. 279–308. 25c.; (r) Nonmarine Cretaceous invertebrates of the San Juan Basin, by T. W. Stanton, pp. 309–326, 15c.; (s) Flora of the Fruitland and Kirtland formations, by F. H. Knowlton, pp. 327–353, 20c.

PROFESSIONAL PAPERS—Continued.

100-A The coal fields of the United States: General introduction, by M. R. Campbell, pp. 1-38. 40c.

101. Geology and paleontology of the Raton Mesa and other regions in Colorado and New Mexico, by W. T. Lee and F. H. Knowlton. 1917. 450 pp. $1.10.

108. Shorter contributions to general geology, 1917. Contains: A comparison of Paleozoic sections in southern New Mexico, by N. H. Darton, pp. 31-55; The Flaxville gravel and its relation to other terrace gravels of the northern Great Plains, by A. J. Collier and W. T. Thom, jr., pp. 179-184.

119. Reptilian faunas of the Torrejon, Puerco, and underlying Upper Cretaceous formations of San Juan County, N. Mex., by C. W. Gilmore. 1920. 71 pp.

122. Copper deposits of the Tyrone district, N. Mex., by Sidney Paige. 1922. 53 pp. 25c.

130. The Laramie flora of the Denver Basin, with a review of the Laramie problem, by F. H. Knowlton. 1922. 175 pp.

131-F. Revision of the flora of the Green River formation, by F. H. Knowlton, pp. 133-182. 10c.

131-G. Fossil plants from the Tertiary lake beds of south-central Colorado, by F. H. Knowlton, pp. 183-197. 10c.

131-H. The fauna of the so-called Dakota formation of northern central Colorado and its equivalent in southeastern Wyoming, by J. B. Reeside, jr., pp. 199-212. 10c.

132. Shorter contributions to general geology, 1923-24. Contains: Relations of the Wasatch and Green River formations in northwestern Colorado and southern Wyoming, with notes on oil shale in the Green River formation, by J. D. Sears and W. H. Bradley, pp. 93-107.

134. Upper Cretaceous and Tertiary formations of the western part of San Juan Basin, Colo. and N. Mex., by J. B. Reeside, jr., and Flora of the Animas formation, by F. H. Knowlton. 1924. 117 pp.

135. The composition of the river and lake waters of the United States, by F. W. Clarke. 1924. 199 pp.

138. Mining in Colorado, by C. W. Henderson. 1926. 263 pp.

148. Geology and ore deposits of the Leadville mining district, Colo., by S. F. Emmons and others. 1927. 368 pp. $2.50.

149. Correlation of geologic formations between east-central Colorado, central Wyoming, and southern Montana, by W. T. Lee. 1927. 80 pp. 50c.

151. The cephalopods of the Eagle sandstone and related formations in the Western Interior of the United States, by J. B. Reeside, jr. 1927. 87 pp. 60c.

154-G. Algae reefs and oolites of the Green River formation, by W. H. Bradley, pp. 203-222. 30c.

154-J. Additions to the flora of the Green River formation, by R. W. Brown, pp. 279-299. 20c.

155. The flora of the Denver and associated formations of Colorado, by F. H. Knowlton. 142 pp. 80c.

158-A. The occurrence and origin of analcite and meerschaum beds in the Green River formation of Utah, Colorado, and Wyoming, by W. H. Bradley, pp. 1-7. 10c.

158-E. The varves and climate of the Green River epoch, by W. H. Bradley, pp. 87-110. 15c.

BULLETINS:

*1. On hypersthene andesite and on triclinic pyroxene in augitic rocks, by Whitman Cross, with a geological sketch of Buffalo Peaks, Colo., by S. F. Emmons. 1883. 42 pp.

*20. Contributions to the mineralogy of the Rocky Mountains, by Whitman Cross and W. F. Hillebrand. 1885. Contains: Minerals from the basalt of Table Mountain, Golden, Colo., pp. 13-89; Minerals from the neighborhood of Pikes Peak, Colo., pp. 40-74; New mineral species from Colorado, pp. 100-109.

*49. Latitudes and longitudes of certain points in Missouri, Kansas, and New Mexico, by R. S. Woodward. 1889. 133 pp.

*66. On a group of volcanic rocks from the Tewan Mountains, N. Mex., and on the occurrence of primary quartz in certain basalts, by J. P. Iddings. 1890. 34 pp.

*93. Some insects of special interest from Florissant, Colo., and other points in the Tertiaries of Colorado and Utah, by S. H. Scudder. 1892. 35 pp.

106. The Colorado formation and its invertebrate fauna, by T. W. Stanton. 1893. 288 pp.

182. A report on the economic geology of the Silverton quadrangle, Colo., by F. L. Ransome. 1901. 265 pp.

185. Results of spirit leveling, 1900-1901, by H. M. Wilson and others. 219 pp.

201. Results of primary triangulation and primary traverse, 1901-2, by H. M. Wilson and others. 164 pp.

213. Contributions to economic geology, 1902. Contains: Reconnaissance examination of the copper deposits at Pearl, Colo., by A. C. Spencer, pp. 163-169.

216. Results of primary triangulation and primary traverse, 1902-3, by S. S. Gannett. 222 pp. 20c.

225. Contributions to economic geology, 1903. Contains: Coal fields of the White Mountain region, N. Mex., by C. A. Fisher, pp. 293-294; Structure of the Boulder oil fields, Colo., with records for the year 1903, by N. M. Fenneman, pp. 383-391.

245. Results of primary triangulation and primary traverse, 1903-4, by S. S. Gannett. 328 pp.

254. Report of progress in the geological resurvey of the Cripple Creek district, Colo., by Waldemar Lindgren and F. L. Ransome. 1904. 36 pp. 5c.

260. Contributions to economic geology, 1904. Contains: Ore deposits of the Ouray district, Colo., by J. D. Irving, pp. 50-77; Ore deposits in the vicinity of Lake City, Colo., by J. D. Irving, pp. 78-84; The geological resurvey of the Cripple Creek district, Colo., by Waldemar Lindgren and F. L. Ransome, pp. 85-98; Preliminary report on ore deposits of the Georgetown mining district, Colo., by J. E. Spurr and G. H. Garrey, pp. 99-120; The Neglected mine and near-by properties, Colo., by W. H. Emmons, pp. 121-127; Copper in the red beds of the Colorado Plateau region, by S. F. Emmons, pp. 221-232; The Florence oil field, Colo., by N. M. Fenneman, pp. 436-440; Zuni salt deposits, N. Mex., by N. H. Darton, pp. 565-566.

262. Contributions to mineralogy from the United States Geological Survey, by F. W. Clarke and others. 1905. Contains: On carnotite and associated vanadiferous minerals in western Colorado, pp. 9-31; Two tellurium minerals from Colorado, pp. 55-57.

265. Geology of the Boulder district, Colo., by N. M. Fenneman. 1905. 101 pp.

276. Results of primary triangulation and primary traverse, 1904-5, by S. S. Gannett. 263 pp. 20c.

285. Contributions to economic geology, 1905. Contains: *(a) Gold and silver—Ore deposits of Bear Creek, near Silverton, Colo., by W. H. Emmons, pp. 25-27, The Hahns Peak gold field, Colo., by H. S. Gale, pp. 28-34, The Idaho Springs mining district, Colo., by J. E. Spurr and G. H. Garrey, pp. 35-40, A reconnaissance of the mineral deposits of New Mexico, by Waldemar Lindgren and L. C. Graton, pp. 74-86; *(b) Copper—The Cashin mine, Colo., by W. H. Emmons, pp. 125-128; *(f) Coal, lignite, and peat—The Yampa coal field, Colo., by N. M. Fenneman and H. S. Gale, pp. 226-239, The Engle coal field, N. Mex., by W. T. Lee, p. 240, The Durango-Gallup coal field of Colorado and New Mexico, by F. C. Schrader, pp. 241-258; *(k) Gypsum and plasters—Gypsum of the Uncompahgre region, Colo., by C. E. Siebenthal, pp. 401-403; (o) Miscellaneous nonmetals—Volcanic ash near Durango, Colo., by L. H. Woolsey, pp. 476-479, 5c.

291. A gazetteer of Colorado, by Henry Gannett. 1906. 185 pp. 15c.

*297. The Yampa coal field, Colo., by N. M. Fenneman and H. S. Gale. 1906. 96 pp.

*302. The areas of the United States, the States, and the Territories, by Henry Gannett. 1906. 9 pp.

*310. Results of primary triangulation and primary traverse, 1905-6, by S. S. Gannett. 1907. 248 pp.

BULLETINS—Continued.

*815. Contributions to economic geology, 1906, Part I. Contains: *(a) Gold and silver—Lake Fork extension of the Silverton mining area, Colo., by L. H. Woolsey, pp. 26–30; *(c) Nickel, uranium, etc.—Carnotite in Rio Blanco County, Colo., by H. S. Gale, pp. 110–117; *(e) Aluminum and bauxite—The Gila River alum deposits, N. Mex., by C. W. Hayes, pp. 215–223; (h) Gypsum, plasters, etc.—Gypsum in northwestern New Mexico, by M. K. Shaler, pp. 260–265, 5c.; *(i) Clays and clay products—Clay deposits of the western part of the Durango-Gallup coal field, N. Mex., by M. K. Shaler and J. H. Gardner, pp. 296–302.

*816. Contributions to economic geology, 1906, Part II. Contains: Coal fields of the Danforth Hills and Grand Hogback, in northwestern Colorado, by H. S. Gale, pp. 264–301; The Book Cliffs coal field, between Grand River, Colo., and Sunnyside, Utah, by G. B. Richardson, pp. 302–320; The Durango coal district, Colo., by J. A. Taff, pp. 321–337; A reconnaissance survey of the western part of the Durango-Gallup coal field of Colorado and New Mexico, by M. K. Shaler, pp. 375–426; The Una del Gato coal field, N. Mex., by M. R. Campbell, pp. 427–430; Coal in the vicinity of Fort Stanton Reservation, N. Mex., by M. R. Campbell, pp. 431–484.

*820. The Downtown district of Leadville, Colo., by S. F. Emmons and J. D. Irving. 1907. 75 pp.

*840. Contributions to economic geology, 1907, Part I. Contains: Gold placer deposits near Lay, Colo., by H. S. Gale, pp. 84–95; Notes on copper deposits in Chaffee, Fremont, and Jefferson Counties, Colo., by Waldemar Lindgren, pp. 157–174; Carnotite and associated minerals in western Routt County, Colo., by H. S. Gale, pp. 257–262; Meerschaum in New Mexico, by D. B. Sterrett, pp. 466–473.

*841. Contributions to economic geology, 1907, Part II. Contains: Coal fields of northwestern Colorado and northeastern Utah, by H. S. Gale, pp. 283–315; The Grand Mesa coal field, Colo., by W. T. Lee, pp. 316–334; The coal field between Gallina and Ratón Spring. N. Mex., in the San Juan coal region, by J. H. Gardner. pp. 335–351; The coal field between Durango, Colo., and Monero, N. Mex., by J. H. Gardner, pp. 352–363; The coal field between Gallup and San Mateo, N. Mex., by J. H. Gardner, pp. 364–378.

*850. Geology of the Rangely oil district, Colo., by H. S. Gale. 1908. 61 pp.

371. Reconnaissance of the Book Cliffs coal field, between Grand River, Colo., and Sunnyside, Utah, by G. B. Richardson. 1909. 54 pp. 20c.

*380. Contributions to economic geology, 1908, Part I. Contains: Notes on the economic geology of southeastern Gunnison County, Colo., by J. M. Hill, pp. 21–40; The Tres Hermanas mining district, N. Mex., by Waldemar Lindgren, pp. 123–128; The Taylor Peak and Whitepine iron-ore deposits, Colo., by E. C. Harder, pp. 188–198; The Hanover iron-ore deposits, N. Mex., by Sidney Paige, pp. 199–214; The Niobrara limestone of northern Colorado as a possible source of Portland cement material, by G. C. Martin, pp. 314–326.

*381. Contributions to economic geology, 1908, Part II. Contains: Coal of the Denver Basin, Colo., by G. C. Martin, pp. 297–306; The South Park coal field, Colo., by C. W. Washburne, pp. 307–316; The Colorado Springs coal field, Colo., by M. I. Goldman, pp. 317–340; The Canon City coal field, Colo., by C. W. Washburne, pp. 341–378; The Trinidad coal field, Colo., by G. B. Richardson, pp. 379–446; Isolated coal fields in Santa Fe and San Miguel Counties, N. Mex., by J. H. Gardner, pp. 447–451; The Carthage coal field, N. Mex., by J. H. Gardner, pp. 452–460; The coal field between San Mateo and Cuba, N. Mex., by J. H. Gardner, pp. 461–473; Development in the Boulder oil field, Colo., by C. W. Washburne, pp. 514–516; The Florence oil field, Colo., by C. W. Washburne, pp. 517–544.

386. Pleistocene geology of the Leadville quadrangle, Colo., by S. R. Capps, jr. 1909. 99 pp. 30c.

*389. The Manzano group of the Rio Grande Valley, N. Mex., by W. T. Lee and G. H. Girty. 1909. 141 pp.

391. The Devonian fauna of the Ouray limestone, by E. M. Kindle. 1909. 60 pp. 25c.

*415. Coal fields of northwestern Colorado and northeastern Utah, by H. S. Gale. 1910. 265 pp.

*427. Manganese deposits of the United States, by E. C. Harder. 1910. 298 pp.

*435. A reconnaissance of parts of northwestern New Mexico and northern Arizona, by N. H. Darton. 1910. 88 pp.

440. Results of triangulation and primary traverse for the years 1906, 1907, and 1908. 688 pp. 50c.

*470. Contributions to economic geology, 1910, Part I. Contains: The economic geology of Carson camp, Colo., by E. S. Larsen, pp. 30–38; The ore deposits near Pinos Altos, N. Mex., by Sidney Paige, pp. 109–125; Metalliferous ore deposits near the Burro Mountains, N. Mex., by Sidney Paige, pp. 131–150; Clay near Calhan, Colo., by G. B. Richardson, pp. 293–296; Gypsum deposit in Eagle County, Colo., by E. F. Burchard, pp. 354–365; Fluorspar near Deming, N. Mex., by N. H. Darton and E. F. Burchard, pp. 533–545.

471. Contributions to economic geology, 1910, Part II. Contains: The coal resources of Gunnison Valley, Colo., by E. G. Woodruff, pp. 565–573; The Tijeras coal field, N. Mex., by W. T. Lee, pp. 575–578; Miscellaneous analyses of coal samples from various fields of the United States, pp. 629–655. $1.00.

478. Geology and ore deposits near Lake City, Colo., by J. D. Irving and Howland Bancroft. 1911. 128 pp. 25c.

496. Results of triangulation and primary traverse. 1909 and 1910. 392 pp. 30c.

*507. The mining districts of the western United States, by J. M. Hill. 1912. 309 pp.

510. Coal fields of Grand Mesa and the West Elk Mountains, Colo., by W. T. Lee. 1912. 237 pp. 30c.

*522. Portland cement materials and industry in the United States, by E. C. Eckel. 1913. 401 pp.

*523. Nitrate deposits, by H. S. Gale. 1912. 36 pp.

*530. Contributions to economic geology, 1911, Part I. Contains: *(e) A preliminary report on the geology and ore deposits of Creede, Colo., by W. H. Emmons and E. S. Larsen, pp. 42–65; (f) Alunite in the San Cristobal quadrangle, Colo., by E. S. Larsen, pp. 179–183, 5c.; *(h) Notes on the vanadium deposits near Placerville, Colo., by F. L. Hess, pp. 142–156, and Vanadium in the Sierra de los Caballos, N. Mex., by F. L. Hess, pp. 157–160; *(l) Graphite near Raton. N. Mex., by W. T. Lee, pp. 371–374, and Mica in Idaho, New Mexico, and Colorado, by D. B. Sterrett, pp. 375–390.

*531. Contributions to economic geology, 1911, Part II. Contains: Geology and petroleum resources of the De Beque oil field, Colo., by E. G. Woodruff, pp. 54–68; The Cerrillos coal field, N. Mex., by W. T. Lee, pp. 285–312; Miscellaneous analyses of coal samples from various fields of the United States, pp. 331–355.

*540. Contributions to economic geology, 1912, Part I. Contains: The Aberdeen granite quarry near Gunnison, Colo., by J. F. Hunter, pp. 359–362; Potash in western saline deposits, by J. H. Hance, pp. 457–469.

*541. Contributions to economic geology, 1912, Part II. Contains: Petroleum near Dayton, N. Mex., by G. B. Richardson, pp. 135–140; Geology and coal resources of the Sierra Blanca coal field, N. Mex., by C. H. Wegemann, pp. 419–452; Analyses of coal samples from various fields of the United States, by M. R. Campbell, pp. 491–526.

551. Results of triangulation and primary traverse, 1911 and 1912. 396 pp. 30c.

GEOLOGIC AND OTHER REPORTS—Continued

BULLETINS—Continued.

565. Results of spirit leveling in Colorado, 1896 to 1914. 192 pp. 15c.
*580. Contributions to economic geology, 1913, Part I. Contains: Notes on the Unaweep copper district, Colo., by B. S. Butler, pp. 19–23; Some cerusite deposits in Custer County, Colo., by J. F. Hunter, pp. 25–37; Some deposits of mica in the United States, by D. B. Sterrett, pp. 65–125.
*581. Contributions to economic geology, 1913, Part II. Contains: Oil shale of northwestern Colorado and northeastern Utah, by E. G. Woodruff and D. T. Day, pp. 1–21.
*583. Colorado ferberite and the wolframite series, by F. L. Hess and W. T. Schaller. 1914. 75 pp.
596. Geology and coal resources of North Park, Colo., by A. L. Beekly. 1915. 121 pp. 15c.
*599. Our mineral reserves—how to make America industrially independent, by G. O. Smith. 1914. 48 pp.
613. Guidebook of the western United States, Part C, The Santa Fe Route, with a side trip to the Grand Canyon of the Colorado, by N. H. Darton and others. 1915. 200 pp. $1.
*618. Geology and underground water of Luna County, N. Mex., by N. H. Darton. 1916. 188 pp.
*620. Contributions to economic geology, 1915, Part I. Contains: Potash in certain copper and gold ores, by B. S. Butler, pp. 227–236; Preliminary report on the economic geology of Gilpin County, Colo., by E. S. Bastin and J. M. Hill, pp. 295–323; The Aztec gold mine, Baldy, N. Mex., by W. T. Lee, pp. 325–380.
*621. Contributions to economic geology, 1915, Part II. Contains: Analyses of coal samples from various parts of the United States, by M. R. Campbell and F. R. Clark, pp. 251–375.
*623. Petroleum withdrawals and restorations affecting the public domain, by M. W. Ball. 1916. 427 pp.
*624. Useful minerals of the United States, by F. C. Schrader and others. 1916. 412 pp.
*625. The enrichment of ore deposits, by W. H. Emmons. 1917. 530 pp.
638. Spirit leveling in New Mexico, 1902 to 1915. 112 pp. 10c.
*641. Contributions to economic geology, 1916, Part II. Contains: Oil shale in northwestern Colorado and adjacent areas, by D. E. Winchester, pp. 139–198.
644–B. Triangulation in Arizona and New Mexico, 1913–1915, pp. 13–24. 5c.
644–D. Triangulation in Colorado, etc., 1913–1915, pp. 85–223. 5c.
652. Tungsten minerals and deposits, by F. L. Hess. 1917. 85 pp. 25c.
*669. Salt resources of the United States, by W. C. Phalen. 1919. 284 pp.
681. The oxidized zinc ores of Leadville, Colo., by G. F. Loughlin. 1918. 91 pp. 20c.
685. Relation of landslides and glacial deposits to reservoir sites in the San Juan Mountains, Colo., by W. W. Atwood. 1918. 38 pp. 10c.
689. Boundaries, areas, geographic centers, and altitudes of the United States and of the several States, by E. M. Douglas. 1923. 284 pp. 50c.
*691. Contributions to economic geology, 1918, Part II. Contains: *(a) The structure of parts of the central Great Plains, by N. H. Darton, pp. 1–26; (k) Coal south of Mancos, Colo., by A. J. Collier, pp. 293–310. 10c.
*697. Gypsum deposits of the United States, by R. W. Stone and others. 1920. 326 pp.
707. Guidebook of the western United States, Part E, The Denver & Rio Grande Western Route, by M. R. Campbell. 1922. 266 pp. $1.
*709. Triangulation and primary traverse, 1916–1918. Contains: Triangulation in New Mexico and Texas, 1915–1917, pp. 739–779; Triangulation in Wyoming and Colorado, 1916–1919, pp. 781–798.
710. Contributions to economic geology, 1919, Part I. Contains: Deposits of manganese ore in New Mexico, by E. L. Jones, jr., pp. 37–60. 30c.
*715. Contributions to economic geology, 1920, Part I. Contains: Some deposits of manganese ore in Colorado, by E. L. Jones, jr., pp. 61–72; The Mogollon district, N. Mex., by H. G. Ferguson, pp. 171–204; Permian salt deposits of the south-central United States, by N. H. Darton, pp. 205–230.
716. Contributions to economic geology, 1920, Part II. Contains: Geology of Alamosa Creek Valley, N. Mex., by D. E. Winchester, pp. 1–15; Coal in the middle and eastern parts of San Juan County, N. Mex., by C. M. Bauer and J. B. Reeside, jr., pp. 155–237. 70c.
718. Geology and ore deposits of the Creede district, Colo., by W. H. Emmons and E. S. Larsen. 1923. 198 pp. 40c.
*726. Contributions to economic geology, 1921, Part II. Contains: Geologic structure of parts of New Mexico, by N. H. Darton, pp. 173–275.
729. Oil shale of the Rocky Mountain region, by D. E. Winchester. 1923. 204 pp. 35c.
735. Contributions to economic geology, 1922, Part I. Contains: Silver enrichment in the San Juan Mountains, Colo., by E. S. Bastin, pp. 65–129. 55c.
740. Mica deposits of the United States, by D. B. Sterrett. 1923. 342 pp. 50c.
*748. The Twentymile Park district of the Yampa coal field, Colo., by M. R. Campbell. 1923. 82 pp.
750–C. Observations on the rich silver ores of Aspen, Colo., by E. S. Bastin, pp. 41–62. 5c.
750–D. New and known minerals from the Utah-Colorado carnotite region, by F. L. Hess, pp. 63–78. 10c.
*751. Contributions to economic geology, 1923–24, Part II. Contains: *(a) Continuity of some oil-bearing sands of Colorado and Wyoming, by W. T. Lee, pp. 1–22; (g) Geology and oil and gas prospects of part of Moffat County, Colo., and southern Sweetwater County, Wyo., by J. D. Sears, pp. 269–319. 20c.
752. Coal resources of the Raton coal field, N. Mex., by W. T. Lee. 1924. 254 pp. 50c.
757. Geology and coal resources of the Axial and Monument Butte quadrangles, Colo., by E. T. Hancock. 1925. 134 pp. 35c.
760. Contributions to the geography of the United States, 1923–24. Contains: Erosion by solution and fill, by W. T. Lee, pp. 107–121. 65c.
767. Geology and coal resources of the Gallup-Zuni Basin, N. Mex., by J. D. Sears. 1925. 53 pp. 50c.
777. Pre-Cambrian rocks of Gunnison River, Colo., by J. F. Hunter. 1925. 94 pp. 30c.
779. Guides to ore in the Leadville district, Colo., by G. F. Loughlin. 1926. 37 pp. 35c.
*780. Contributions to economic geology, 1925, Part I. Contains: Geology of a part of western Texas and southeastern New Mexico, with special reference to salt and potash, by H. W. Hoots, pp. 33–126.
785–A. Recent developments in the Aspen district, Colo., by Adolph Knopf, pp. 1–28. 10c.
787. Geology and ore deposits of the Mogollon mining district, N. Mex., by H. G. Ferguson. 1927. 100 pp. 65c.
790–A. Pedestal rocks formed by differential erosion and Channel erosion of the Rio Salado, N. Mex., by Kirk Bryan, pp. 1–19. 10c.
794. "Red Beds" and associated formations in New Mexico, by N. H. Darton. 1928. 221 pp. $1.30.
796–B. Geology and oil and gas prospects of northeastern Colorado, by K. F. Mather, James Gilluly, and R. G. Lusk, pp. 65–124. 20c.
811–B. Recent mining developments in the Creede district, Colo., by E. S. Larsen, pp. 89–112. 10c.
812–C. Geology and coal resources of the Meeker quadrangle, Colorado, by E. T. Hancock and J. B. Eby, pp. 191–242. 30c.
822–B. The Granby anticline, Grand County, Colo., by T. S. Lovering, pp. 71–76. 5c.

6

MAPS OF THE UNITED STATES

A wall map, 55 by 85 inches, in two sheets, on a scale of 37 miles to 1 inch, without contours, showing coal fields. 1917. Price, $1; if included in wholesale orders, 60 cents

A wall map, 49 by 76 inches, in two sheets, on a scale of 40 miles to 1 inch, either with or without contours. Price, 75 cents; if included in wholesale orders, 45 cents. 1922.

A wall map, same size and scale as next above, without contours, showing producing coal districts.

A wall map, 40 by 62 inches, on a scale of 50 miles to 1 inch, on which is indicated by depth of brown and blue colors the relative height of the land and the depth of the sea. The position of the principal cities and the boundaries of the States are shown. Price, 75 cents; in lots of 10 or more, 50 cents.

A map, 18 by 28 inches, on a scale of 110 miles to 1 inch, either with or without contours. Price, 15 cents; if included in wholesale orders, 9 cents.

A relief or hypsometric map, same size, scale, and price as next above; altitudes indicated by colors.

A base map, 11 by 16 inches, on a scale of 190 miles to 1 inch. Price, 5 cents; if included in whole-...

GEOLOGIC AND OTHER REPORTS—Continued

WATER-SUPPLY PAPERS:

- *3. Sewage irrigation, by G. W. Rafter. 1897. 100 pp.
- *9. Irrigation near Greeley, Colo., by David Boyd. 1897. 90 pp.
- *10. Irrigation in Mesilla Valley, N. Mex. by F. C. Barker. 1898. 51 pp.
- *22. Sewage irrigation, Part II, by G. W. Rafter. 1899. 100 pp.
- *44. Profiles of rivers in the United States, by Henry Gannett. 1901. 100 pp.
- 74. Water resources of the State of Colorado, by A. L. Fellows. 1902. 151 pp. 25c.
- *93. Proceedings of first conference of engineers of the Reclamation Service, compiled by F. H. Newell. 1904. 361 pp.
- 123. Geology and underground water conditions of the Jornada del Muerto, N. Mex., by C. R. Keyes. 1905. 42 pp. 15c.
- *141. Observations on the ground waters of Rio Grande Valley, by C. S. Slichter. 1905. 83 pp.
- 147. Destructive floods in the United States in 1904, by E. C. Murphy and others. 1905. 206 pp. 15c.
- 149. Preliminary list of deep borings in the United States, by N. H. Darton. 1905. 175 pp. 10c.
- *158. Preliminary report on the geology and underground waters of the Roswell artesian area, N. Mex., by C. A. Fisher. 1906. 29 pp.
- *162. Destructive floods in the United States in 1905, by E. C. Murphy and others. 1906. 105 pp.
- *184. The underflow of the South Platte Valley, by C. S. Slichter and H. C. Wolff. 1906. 42 pp.
- *188. Water resources of the Rio Grande Valley in New Mexico and their development, by W. T. Lee. 1907. 59 pp.
- *240. Geology and water resources of the San Luis Valley, Colo., by C. E. Siebenthal. 1910. 128 pp.
- *260. Preliminary report on the ground waters of Estancia Valley, N. Mex., by O. E. Meinzer. 1910. 33 pp.
- *274. Some stream waters of the western United States, by Herman Stabler. 1911. 188 pp.
- *275. Geology and water resources of Estancia Valley, N. Mex., by O. E. Meinzer. 1911. 89 pp.
- 340. Stream-gaging stations and publications relating to water resources, 1885–1913, by B. D. Wood. 1916. 195 pp. 15c.
- *343. Geology and water resources of Tularosa Basin, N. Mex., by O. E. Meinzer. 1915. 317 pp.
- 345. Contributions to the hydrology of the United States, 1914. 30c. Contains: (b) Ground water for irrigation in the vicinity of Enid, Okla., by A. T. Schwennesen, pp. 11–23, 5c.; *(c) Underground water of Luna County, N. Mex., by N. H. Darton, pp. 25–40.
- 364. Water analyses from the laboratory of the United States Geological Survey, by F. W. Clarke. 1914. 40 pp. 5c.
- *380. The Navajo country—a geographic and hydrographic reconnaissance of parts of Arizona, New Mexico, and Utah. by H. E. Gregory. 1916. 219 pp.
- 395. Colorado River and its utilization, by E. C. La Rue. 1916. 231 pp. 50c.
- 396. Profile surveys in the Colorado River Basin in Wyoming, Utah, Colorado. and New Mexico. 1917. 6 pp. 50c.
- 398. Ground water in San Joaquin Valley, Calif., by W. C. Mendenhall and others. 1916. 310 pp. 25c.
- 421. Profile surveys in 1915 along Rio Grande, Pecos River, and Mora River, N. Mex. 1916. 11 pp. 15c.
- 422. Ground water in the Animas, Playas, Hachita, and San Luis Basins, N. Mex., by A. T. Schwennesen. 1918. 152 pp. 20c.
- *425. Contributions to the hydrology of the United States, 1917. Contains: Ground water in San Simon Valley, Ariz. and N. Mex., by A. T. Schwennesen, pp. 1–35.
- *427. Bibliography and index of the publications of the United States Geological Survey relating to ground water, by O. E. Meinzer. 1918. 169 pp.
- 487. The Arkansas River flood of June 3–5, 1921, by Robert Follansbee and E. E. Jones. 1922. 44 pp. 10c.
- *489. The occurrence of ground water in the United States, by O. E. Meinzer. 1923. 321 pp.
- 556. Water power and flood control of Colorado River below Green River, Utah, by E. C. LaRue. 1925. 176 pp. $1.
- 557. Large springs in the United States, by O. E. Meinzer. 1927. 94 pp. 30c.
- 558. Preliminary index to river surveys made by the United States Geological Survey and other agencies, by B. E. Jones and R. O. Helland. 1926. 108 pp. 25c.
- 560–B. Chemical character of ground waters of the northern Great Plains, by H. B. Riffenburg, pp. 31–52. 5c.
- 580–A. Geology of No. 3 reservoir site of the Carlsbad irrigation project, N. Mex., with respect to water-tightness, by O. E. Meinzer and others, pp. 1–39. 15c.
- 596–B. Quality of water of Colorado River in 1925–26, by W. D. Collins and C. S. Howard, pp. 33–43. 10c.
- 617. Upper Colorado River and its utilization, by Robert Follansbee. 394 pp. 65c.
- 618. The Green River and its utilization, by Ralf R. Woolley. 456 pp. $1.25.
- 638–A. Quality of water of the Colorado River in 1926–1928, by C. S. Howard, pp. 1–14. 5c.
- 638–B. Suspended matter in the Colorado River in 1925–1928, by C. S. Howard, pp. 15–44. 10c.

Stream measurements in the years mentioned:

Year	Water-Supply Paper	Price (cents)	Year	Water-Supply Paper	Price (cents)	Year	Water-Supply Paper	Price (cents)	Year	Water-Supply Paper	Price (cents)
1897	*15		1907–8	246	30	1914	386	30	1921	526	30
	*16			247	15		387	5		527	5
1898	28	10		248	20		388	10		528	10
1899	*37			249	25		389	15		529	25
	*38		1909	266	25	1915	406	20	1922	546	35
1900	49	10		267	10		407	10		547	15
	50	10		288	15		408	10		548	15
1901	66	10		269	25		409	20		549	25
	75	25	1910	286	25	1916	436	20	1923	566	50
1902	84	10		287	10		437	10		567	15
	85	15		288	15		438	10		568	30

GEOLOGIC AND OTHER REPORTS

The following reports relate to Colorado and New Mexico but are not parts of the topographic or geologic atlas. An asterisk (*) indicates that the report is out of print, but many such reports are available for consultation in certain libraries. (See list on p. 8.) The publications for which the price is stated are sold by the Superintendent of Documents, Government Printing Office, Washington, D. C. Remittance to that official should be made by postal money order, express order, or check; postage stamps will not be accepted.

ANNUAL REPORTS:

*Second, 1880–81. Contains: Abstract of report on geology and mining industry of Leadville, Colo., by S. F. Emmons. pp. 201–290.

Sixth. 1884–85. Contains: Mount Taylor and the Zuni Plateau, by C. E. Dutton, pp. 105–198. $2.

Eighth, 1886–87. I contains: The fossil butterflies of Florissant, Colo., by S. H. Scudder, pp. 433–474. $1.50.Part

Ninth, 1887–88. Contains: On the geology and physiography of a portion of northwestern Colorado and adjacent parts of Utah and Wyoming, by C. A. White, pp. 677–712. $2.

Thirteenth, 1891–92. Part III contains: Report upon the construction of topographic maps and the selection and survey of reservoir sites in the hydrographic basin of Arkansas River, Colo., by A. H. Thompson, pp. 429–444. $1.85.

Fourteenth, 1892–93. *Part II contains: The laccolitic mountain groups of Colorado, Utah, and Arizona, by Whitman Cross, pp. 157–241.

Sixteenth, 1894–95. Part II contains: Geology and mining industries of the Cripple Creek district, Colo., by Whitman Cross and R. A. F. Penrose, jr., pp. 1–209; Water resources of a portion of the Great Plains. by Robert Hay, pp. 535–585. $1.25.

Seventeenth, 1895–96. *Part II contains: Geology of Silver Cliff and the Rosita Hills, Colo., by Whitman Cross, pp. 263–403; The mines of Custer County, Colo., by S. F. Emmons, pp. 405–472; The underground water of the Arkansas Valley in eastern Colorado, by G. K. Gilbert, pp. 551–601.

Eighteenth. 1896–97. Part I contains: Triangulation and spirit leveling, by H. M. Wilson and others, pp. 131–422; $1. *Part III contains: Preliminary report on the mining industries of the Telluride quadrangle, Colo., by C. W. Purington, pp. 745–850.

Nineteenth, 1897–98. Part I contains: Triangulation and spirit leveling, by H. M. Wilson and others, pp. 145–408. $1.

Twentieth, 1898–99. Part I contains: Triangulation and spirit leveling, by H. M. Wilson and others. pp. 211–530; $1. Part II contains: Devonian fossils from southwestern Colorado; The fauna of the Ouray limestone, by G. H. Girty, pp. 25–81; $2.50. Part V contains: Pikes Peak, Plum Creek, and South Platte Reserves, by J. G. Jack, pp. 39–115; White River Plateau Timber Land Reserve, by G. B. Sudworth, pp. 117–179, and Battlement Mesa Forest Reserve, by G. B. Sudworth, pp. 181–243; $2.80.

Twenty-first, 1899–1900. *Part II contains: Geology of the Rico Mountains, Colo., by Whitman Cross and A. C. Spencer, pp. 7–165. Part IV contains: The High Plains and their utilization, by W. D. Johnson, pp. 601–741; $2.85.

Twenty-second, 1900–1901. *Part I contains: The asphalt and bituminous rock deposits of the United States, by G. H. Eldridge, pp. 209–452. *Part II contains: The ore deposits of the Rico Mountains, Colo., by F. L. Ransome, pp. 229–398. *Part III contains: The Rocky Mountain coal fields, by L. S. Storrs, pp. 415–471. Part IV contains: The High Plains and their utilization, by W. D. Johnson, pp. 631–669; $2.20.

MONOGRAPHS:

*12. Geolgy and mining industry of Leadville, Colo., by S. F. Emmons. 1886. 770 pp.

*27. Geology of the Denver Basin in Colorado, by S. F. Emmons and others. 1896. 556 pp.

31. Geology of the Aspen mining district, Colo., by J. E. Spurr. 1898. 260 pp. $3.60.

35. The later extinct floras of North America, by J. S. Newberry. 1898. 295 pp. $1.25.

40. Adephagous and clavicorn Coleoptera from the Tertiary deposits at Florissant, Colo., by S. H. Scudder. 1900. 148 pp. 80c.

44. Pseudoceratites of the Cretaceous, by Alpheus Hyatt, 1903. 351 pp. $1.

49. The Ceratopsia, by J. B. Hatcher. 1907. 300 pp. $1.50.

51. Cambrian Brachiopoda, by C. D. Walcott. 1912. In two parts. Part I, 872 pp.; Part II, 363 pp. $4.

54. The Mesozoic and Cenozoic Echinodermata of the United States, by W. B. Clark and M. W. Twitchell. 1915. 341 pp. $1.50.

PROFESSIONAL PAPERS:

*16. The Carboniferous formations and faunas of Colorado, by G. H. Girty. 1903. 546 pp.

*32. Preliminary report on the geology and underground water resources of the central Great Plains, by N. H. Darton. 1905. 433 pp.

*33. Forest conditions in the Lincoln Forest Reserve, N. Mex., by F. G. Plummer and M. G. Gowsell. 1904. 47 pp.

*39. Forest conditions in the Gila River Forest Reserve, N. Nex., by T. F. Rixon. 1905. 89 pp.

*52. Geology and underground waters of the Arkansas Valley in eastern Colorado, by N. H. Darton. 1906. 90 pp.

*54. Geology and gold deposits of the Cripple Creek district, Colo., by Waldemar Lindgren and F. L. Ransome. 1906. 516 pp.

*63. Economic geology of the Georgetown quadrangle (together with the Empire district), Colo., by J. E. Spurr and G. H. Garrey. 1908. 422 pp.

67. Landslides in the San Juan Mountains, Colo., by Ernest Howe. 1909. 58 pp. 25c.

*68. The ore deposits of New Mexico, by Waldemar Lindgren and others. 1910. 361 pp.

75. Geology and ore deposits of the Breckenridge district, Colo., by F. L. Ransome 1911 187 pp. 75c.

*90. Shorter contributions to general geology, 1914. Contains *(a) Geology of the pitchblende ores of Colorado, by E. S. Bastin. pp. 1–5; *(c) Dike rocks of the Apishapa quadrangle, Colo., by Whitman Cross, pp. 17–31; (r) Contributions to the stratigraphy of southwestern Colorado, by Whitman Cross and E. S. Larsen, pp. 39–50, 10c.; *(k) The history of a portion of Yampa River, Colo.. and its possible bearing on that of Green River, by E. T. Hancock, pp. 183–189.

*93. Geology of the Navajo country—a reconnaissance of parts of Arizona, New Mexico, and Utah, by H. E. Gregory. 1917. 161 pp.

*94. Economic geology of Gilpin County and adjacent parts of Clear Creek and Boulder Counties, Colo, by E. S. Bastin and J. M. Hill. 1917. 379 pp.

*95. Shorter contributions to general geology, 1915. Contains: (b) Eocene glacial deposits in southwestern Colorado, by W. W. Atwood, pp. 13–26, 15c.: *(c) Relation of the Cretaceous formations to the Rocky Mountains in Colorado and New Mexico, by W. T. Lee, pp. 27–58.

*98. Shorter contributions to general geology, 1916. Contains: *(h) The flora of the Fox Hills sandstone, by F. H. Knowlton, pp. 85–93; (p) Stratigraphy of a part of the Chaco River Valley, N. Mex., by C. M. Bauer, pp. 271–278, 15c.; (q) Vertebrate faunas of the Ojo Alamo. Kirtland, and Fruitland formations, by C. W. Gilmore, pp. 279–308. 25c.; (r) Nonmarine Cretaceous invertebrates of the San Juan Basin, by T. W. Stanton, pp. 309–326, 15c.; (s) Flora of the Fruitland and Kirtland formations, by F. H. Knowlton, pp. 327–353, 20c.

8

Libraries.—Many of the publications listed above may be consulted in the following libraries in Colorado and New Mexico.

COLORADO

BOULDER:
  University of Colorado.
COLORADO SPRINGS:
  Colorado College.
DENVER:
  Colorado State.
  Museum of Natural History.

DENVER—Continued.
  Public.
  Regis College.
  State Historical & Natural History Society.
  State Mining Bureau.
  University of Denver.

FORT COLLINS:
  Experiment Station.
  State Agricultural College.
GOLDEN:
  State School of Mines.
PUEBLO:
  McClelland Public.

NEW MEXICO

ALBUQUERQUE:
  University of New Mexico.
EAST LAS VEGAS:
  New Mexico Normal University.

SANTA FE:
  State.
SILVER CITY:
  New Mexico State Teachers College.

SOCORRO:
  School of Mines.
STATE COLLEGE:
  New Mexico College of Agriculture and Mechanical Arts.

SPECIAL MAPS AND SHEETS

[Measurements are approximate]

Alamo National Forest, N. Mex. This map shows the Alamo National Forest, in Otero and Chaves Counties, N. Mex. Limiting p e , 32° 30′ and 33° 30′. Limiting meridians, 105° and 106°. Size, 16½ by 20 inches. Scale, 1:250,000, or about 4 miles to 1 inch. Contour interval, 200 feet. Price, 10 cents; if included in wholesale orders, 6 cents.

Aspen special, Colo. This map shows a portion of the Aspen quadrangle, in the vicinity of Aspen, on a large scale. Size, 18 by 22 inches. Scale, 1:9,600, or 800 feet to 1 inch. Contour interval, 25 feet. Price, 10 cents; if included in wholesale orders, 6 cents.

Breckenridge special, Colo. This map shows a portion of the Leadville quadrangle, in the vicinity of Breckenridge, on a large scale. Limiting parallels, 39° 27′ and 39° 33′. Limiting meridians, 105° 56′ 58″ and 106° 04′. Size, 19 by 22 inches. Scale, 1:24,000, or 2,000 feet to 1 inch. Contour interval, 50 feet. Price, 10 cents; if included in wholesale orders, 6 cents.

Central City special, Colo. This map shows portions of the Central City and Blackhawk quadrangles, in the vicinity of Central City, on a large scale. Limiting parallels, 39° 46′ 23″ and 39° 49′ 06″. Limiting meridians, 105° 29′ 15″ and 105° 32′ 46″. Size, 19 by 21 inches. Scale, 1:12,000, or 1,000 feet to 1 inch. Contour interval, 50 feet. Price, 10 cents; if included in wholesale orders, 6 cents.

Colorado (State). This map, which is in two colors, shows county boundaries, location and names of all towns, most of the smaller settlements, and the railroads (in black), also the rivers and many of the smaller streams and water features (in blue); it does not show contours. Size, 44 by 53 inches. Scale, 1:500,000, or about 8 miles to 1 inch. Price, 25 cents; if included in wholesale orders, 15 cents.

Creede and vicinity, Colo. This map shows the mining district adjacent to Creede on a large scale. Limiting parallels, 37° 50′ and 37° 55′. Limiting meridians, 106° 54′ and 106° 59′. Size, 16½ by 21 inches. Scale, 1:24,000, or 2,000 feet to 1 inch. Contour interval, 50 feet. Price, 10 cents; if included in wholesale orders, 6 cents.

Cripple Creek, Colo. This map shows a portion of the Pikes Peak quadrangle, in the vicinity of Cripple Creek, on a large scale. Limiting parallels, 38° 41′ 10″ and 38° 47′ 10″. Limiting meridians, 105° 05′ 45″ and 105° 12′ 21″. Size, 16½ by 20 inches. Scale, 1:25,000, or about 2,100 feet to 1 inch. Contour interval, 50 feet. Price, 10 cents; if included in wholesale orders, 6 cents.

De Beque oil field, Colo. This map shows an area 12 miles square on Grand River in Garfield and Mesa Counties, Colo." Limiting parallels, 39° 18′ 58″ and 39° 28′ 20″. Limiting meridians, 108° 09′ 27″ and 108° 22′ 47″. Size, 16½ by 20 inches. Scale, 1:62,500, or about 1 mile to 1 inch. Contour interval, 50 feet. Price, 10 cents; if included in wholesale orders, 6 cents.

Denver, Colo. This map, which is double the standard size, is counted as two sheets in the text of this circular. Limiting parallels, 39° 30′ and 40°. Limiting meridians, 104° 30′ and 105° 20′ (area in southwest corner unsurveyed). Size, 20 by 33 inches. Scale, 1:125,000, or about 2 miles to 1 inch. Contour interval, 50 and 100 feet. Price, 20 cents; if included in wholesale orders, 12 cents.

Denver Mountain Parks, Colo. This map shows the mountain playground west of Denver. Limiting parallels, 39° 30′ and 39° 45′. Limiting meridians, 105° 10′ and 105° 40′. Size, 22 by 30 inches. Scale, 1:62,500, or about 1 mile to 1 inch. Contour interval, 100 feet. Price, 15 cents; if included in wholesale orders, 9 cents.

Fort Bayard special, N. Mex. This map shows the Fort Bayard Military Reservation on a large scale. Limiting parallels, 32° 46′ 42″ and 32° 51′ 53″. Limiting meridians, 108° 07′ 47″ and 108° 11′ 04″. Size, 22 by 35 inches. Scale, 1:12,000, or 1,000 feet to 1 inch. Contour interval, 10 feet. Price, 20 cents; if included in wholesale orders, 12 cents.

Green River, Wyo.-Utah. Plan and profile of Green River from Green River, Utah, to Green River, Wyo. Scale, 1:31,680, or one-half mile to 1 inch. Contour interval on land 20 feet, on river surface 5 feet. Vertical scale of profiles, 20 feet=1 inch. Size, 21 by 27 inches. 16 sheets (10 plans, 6 profiles). Price, 10 cents a sheet; if included in wholesale orders, 6 cents.

Idaho Springs special, Colo. This map shows portions of the Central City and Blackhawk quadrangles, in the vicinity of Idaho Springs, on a large scale. Limiting parallels, 39° 44′ 30″ and 39° 46′ 23″. Limiting meridians, 105° 29′ 50″ and 105° 33′ 30″. Size, 16½ by 20 inches. Scale, 1:12,000, or 1,000 feet to 1 inch. Contour interval, 50 feet. Price, 10 cents; if included in wholesale orders, 6 cents.

Leadville mining district, Colo. This map shows the Leadville ore belt, in Lake County, on a large scale. Limiting parallels, 39° 13′ 50″ and 39° 16′. Limiting meridians, 106° 13′ and 106° 19′ 47″. Size, 20½ by 48 inches. Scale, 1:9,600, or 800 feet to 1 inch. Contour interval, 25 feet. Price, 25 cents; if included in wholesale orders, 15 cents.

Magdalena district, N. Mex. This map shows the Magdalena mining district, in Socorro County, on a large scale. Limiting parallels, 34° 03′ 15″ and 34° 07′. Limiting meridians, 107° 10′ 45″ and 107° 13′. Size, 16½ by 36 inches. Scale, 1:12,000, or 1,000 feet to 1 inch. Contour interval, 25 feet. Price, 10 cents; if included in wholesale orders, 6 cents.

Manitou, Colo. This map shows portions of the Pikes Peak and Colorado Springs quadrangles, in the vicinity of Pikes Peak and Colorado Springs, on a large scale. Limiting parallels, 38° 43′ 30″ and 38° 54′. Limiting meridians, 104° 47′ 30″ and 105° 05′ 45″. Size, 18 by 23 inches. Scale, 1:48,000, or 4,000 feet to 1 inch. Contour interval, 50 feet. Price, 20 cents; if included in wholesale orders, 12 cents.

Mesa Verde National Park, Colo. This map shows the Mesa Verde National Park, in Montezuma County, Colo., on a large scale. Limiting parallels, 37° 09′ 18″ and 37° 21′. Limiting meridians, 108° 15′ and 108° 37′ 30″. Size, 31 by 46 inches. Scale, 1:31,250, or about one-half mile to 1 inch. Contour interval, 25 feet. Price, 20 cents; if included in wholesale orders, 12 cents.

8

[Measurements are approximate]

New Mexico (geologic). This map is printed in two parts, each 30 by 51 inches. Scale, 1:500,000, or 8 miles to 1 inch. geologic formations printed in colors, having 22 pattern distinctions. Base has drainage printed in blue, culture in black, and contours in brown. Price, $1.50; if included in whole-sale orders, 90 cents.

New Mexico (State). This map, which is in two colors, shows county boundaries, location and names of all towns, most of the smaller settlements, and the railroads (in black), also the rivers and most of the smaller streams and water features (in blue); it does not show contours. Published in two sheets; size of each sheet, 28 by 48 inches. Scale, 1:500,000, or about 8 miles to 1 inch. Price, 25 cents; if included in wholesale orders, 15 cents.

New Mexico (State). This map shows by brown contour lines the diversified configuration of the surface of New Mexico. It is the first published contour map of New Mexico. Published in two sheets; size of each sheet, 29 by 49 inches. Scale, 1:500,000, or about 8 miles to 1 inch. Contour interval, 100 meters (328 feet). Price, 75 cents; if included in wholesale orders, 45 cents.

Rico district, Colo. This map shows portions of the Rico and Engineer Mountain quadrangles, in the vicinity of Rico, on a large scale. Limiting parallels, 37° 40′ and 37° 44′ 39″. Limiting meridians, 107° 58′ 37″ and 108° 05′ 39′. Size, 17 by 22 inches. Scale, 1:23,680, or about 2,000 feet to 1 inch. Contour interval, 50 feet. Price, 10 cents; if included in wholesale orders, 6 cents.

River survey maps (advance sheets). The following advance sheets are now available for distribution but are useful chiefly to engineers. Price, 50 cents a sheet; no wholesale rate. (See map for river surveys that are contained in water-supply papers.)

ARKANSAS RIVER AND TRIBUTARIES, COLO.: Three sheets (plan) of Arkansas River from Canon City to sec. 15, T. 8 S., R. 79 W.; includes Twin Lakes and Tennessee Fork to sec. 22, T. 8 S., R. 80 W. Scales, 1:31,680 and 1:63,360. Contour interval, 50 feet. Also one sheet (plan) each of Halfmoon Creek to sec. 19, T. 10 S., R. 81 W.; Lake Creek to sec. 3, T. 11 S., R. 82 W.; Cottonwood Creek to sec. 8, T. 14 S., R. 80 W.; Chalk Creek to Hancock; South Fork of Arkansas River to Monarch; Grape Creek to sec. 36, T. 21 S., R. 73 W.; Beaver Creek from sec. 16, T. 17 S., R. 68 W., to forks, including West Fork to Shagway Reservoir and East Fork to sec. 15, T. 16 S., R. 68 W. Scale, 1:31,680. Contour interval, 50 feet.

BLUE RIVER, COLO.: Two sheets (plan) of Blue River from mouth to Breckinridge, including Green Mountain reservoir site. Scale, 1:31,680. Contour interval, 50 feet.

BOULDER CREEK, COLO.: One sheet (plan) of Middle Boulder Creek from Boulder to Hessie and South Boulder Creek from sec. 26, T. 1 S., R. 71 W., to sec. 6, T. 2 S., R. 73 W. Scale, 1:63,360. Contour interval, 50 feet.

CACHE LA POUDRE RIVER, COLO.: One sheet (plan) of Cache La Poudre River from sec. 33, T. 9 N., R. 70 W., to sec. 33, T. 8 N., R. 75 W. Scale, 1:63,360. Contour interval, 50 feet.

CLEAR CREEK, COLO.: One sheet (plan) of Clear Creek from sec. 36, T. 3 S., R. 71 W., to sec. 17, T. 4 S., R. 75 W., including North Fork, Middle Fork, South Fork, and Chicago Creek. Scale, 1:63,360. Contour interval, 50 feet.

DOLORES RIVER, COLO.: One sheet (plan) of Dolores River from mouth to Paradox Valley. Scale, 1:31,680. Contour interval, 50 feet.

EAGLE RIVER, COLO.: One sheet (plan) of Eagle River from mouth to sec. 30, T. 6 S., R. 80 W. Scale, 1:31,680. Contour interval, 50 feet.

GUNNISON RIVER, COLO.: Three sheets (plan and profile) of Gunnison River from Cimarron Creek to Gunnison. Scale, 1:31,680. No contour lines.

ROARING FORK, COLO.: One sheet (plan) of Roaring Fork from mouth to Snowmass. Scale, 1:31,680. Contour interval, 50 feet.

ST. VRAIN CREEK, COLO.: One sheet (plan) including North Fork of St. Vrain Creek from Lyons to sec. 19, T. 3 N., R. 72 W.; Middle Fork of St. Vrain Creek from Lyons to sec. 18, T. 2 N., R. 72 W.; South Fork to sec. 30, T. 2 N., R. 72 W.; Lefthand Creek from sec. 24, T. 2 N., R. 71 W., to sec. 10, T. 1 N., R. 72 W.; James Creek to sec. 31, T. 2 N., R. 72 W. Scale, 1:63,360. Contour interval, 50 feet.

SAN MIGUEL RIVER, COLO.: One sheet (plan) of San Miguel River from mouth to Sawpit. Scale, 1:31,680. Contour interval, 50 feet.

SAN FRANCISCO RIVER, N. MEX.: One sheet (plan) of Redrock reservoir site on Gila River and Alma reservoir site on San Francisco River. Scale, 1:24,000. Contour interval, 10 feet.

SOUTH PLATTE RIVER, COLO.: Five sheets (plan) of South Platte River from sec. 33, T. 6 S., R. 69 W., to sec. 21, T. 12 S., R. 76 W.; includes North Fork to Webster and Geneva Creek to sec. 13, T. 6 S., R. 75 W. Scale, 1:31,680. Contour interval, 50 feet.

THOMPSON RIVER, COLO.: One sheet (plan) of Thompson River from sec. 12, T. 5 N., R. 70 W., to east boundary of Rocky Mountain National Park. Scale, 1:63,360. Contour interval, 50 feet.

Santa Rita special, N. Mex. This map shows a portion of the Silver City quadrangle, in the vicinity of Santa Rita, on a large scale. Limiting parallels, 32° 46′ and 32° 52′. Limiting meridians, 108° 03′ and 108° 09′. Size, 16¼ by 20 inches. Scale, 1:24,000, or 2,000 feet to 1 inch. Contour interval, 20 feet. Price, 10 cents; if included in wholesale orders, 6 cents.

Silver Plume special, Colo. This map shows a portion of the Georgetown quadrangle, in the vicinity of Silver Plume, on a large scale. Limiting parallels, 39° 41′ 33″ and 39° 42′ 26″. Limiting meridians, 105° 42′ 35″ and 105° 45′. Size, 8½ by 14 inches. Scale, 1:12,000, or 1,000 feet to 1 inch. Contour interval, 50 feet. Price, 10 cents; if included in wholesale orders, 6 cents.

Tenmile district, Colo. This map shows a portion of the Leadville quadrangle, in the vicinity of Kokomo, on a large scale. Limiting parallels, 39° 22′ 57″ and 39° 30′ 25″. Limiting meridians, 106° 08′ and 106° 16′ 08″. Size, 18½ by 21¼ inches. Scale, 1:31,680, or one-half mile to 1 inch. Contour interval, 100 feet. Price, 10 cents; if included in wholesale orders, 6 cents.

Tyrone district, N. Mex. This map shows the Tyrone mining district in Grant County on a large scale. Limiting parallels, 32° 35′ 30″ and 32° 45′ 05″. Limiting meridians, 108° 20′ 80″ and 108° 26′ 30″. Size, 17 by 20 inches. Scale, 1:24,000, or 2,000 feet to 1 inch. Contour interval, 25 feet. Price, 10 cents; if included in wholesale orders, 6 cents.

Yampa River, Colo. Plan and profile of Yampa River from Green River to Morgan Gulch. Scale, 1:31,680, or one-half mile to 1 inch. Contour interval on land 20 feet, on river surface 5 feet. Vertical scale of profiles, 20 feet=1 inch. Size, 21 by 27 inches. 5 sheets (3 plans, 2 profiles). Price, 10 cents a sheet; if included in wholesale orders, 6 cents.

North America. This map does not show contours. Size, 29 by 38 inches. Scale, 1:10,000,000 or about 158 miles to 1 inch. Price, 40 cents; if included in wholesale orders, 24 cents.

Sheet of standard symbols. Shows symbols and abbreviations adopted by the Board of Surveys and Maps of the United States Government for use on Government maps; printed in five colors. Size, 18 by 30 inches. Price, 40 cents; if included in wholesale orders, 24 cents.

GEOLOGIC AND OTHER REPORTS—Continued

WATER-SUPPLY PAPERS:
*8. Sewage irrigation, by G. W. Rafter. 1897. 100 pp.
*9. Irrigation near Greeley, Colo., by David Boyd. 1897. 90 pp.
*10. Irrigation in Mesilla Valley, N. Mex., by F. C. Barker. 1898. 51 pp.
*22. Sewage irrigation, Part II, by G. W. Rafter. 1899. 100 pp.
*44. Profiles of rivers in the United States, by Henry Gannet. 1901. 100 pp.
*74. Water resources of the State of Colorado, by A. L. Fellows. 1902. 151 pp. 25c.
*86. Proceedings of first conference of engineers of the Reclamation Service, compiled by F. H. Newell. 1904. 361 pp.
123. Geology and underground water conditions of the Jornada del Muerto, N. Mex., by C. R. Keyes. 1905. 42 pp. 15c.
*141. Observations on the ground waters of Rio Grande Valley, by C. S. Slichter. 1905. 83 pp.
147. Destructive floods in the United States in 1904, by E. C. Murphy and others. 1905. 206 pp. 15c.
149. Preliminary list of deep borings in the United States, by N. H. Darton. 1905. 175 pp. 10c.
*158. Preliminary report on the geology and underground waters of the Roswell artesian area, N. Mex., by C. A. Fisher. 1906. 29 pp.

MAPS OF THE UNITED STATES

A wall map, 55 by 85 inches, in two sheets, on a scale of 37 miles to 1 inch, without contours, showing coal fields. 1917. Price, $1; if included in wholesale orders, 60 cents

A wall map, 49 by 76 inches, in two sheets, on a scale of 40 miles to 1 inch, either with or without contours. Price, 60 cents; if included in wholesale orders, 36 cents.

A wall map, same size and scale as next above, without contours, showing producing coal districts. 1922. Price, 75 cents; if included in wholesale orders, 45 cents.

A wall map, 40 by 62 inches, on a scale of 50 miles to 1 inch, on which is indicated by depth of brown and blue colors the relative height of the land and the depth of the sea. The position of the principal cities and the boundaries of the States are shown. Price, 75 cents; in lots of 10 or more, 50 cents.

A map, 18 by 28 inches, on a scale of 110 miles to 1 inch, either with or without contours. Price, 15 cents; if included in wholesale orders, 9 cents.

A relief or hypsometric map, same size, scale, and price as next above; altitudes indicated by colors.

A base map, 11 by 16 inches, on a scale of 190 miles to 1 inch. Price, 5 cents; if included in wholesale orders, 3 cents.

A base map, 8¼ by 12 inches, on a scale of 260 miles to 1 inch. Price, 1 cent; if included in wholesale orders, five for 3 cents.

A map, 28 by 31 inches, on a scale of 110 miles to 1 inch, without contours, showing the physical divisions. Price, 10 cents; if included in wholesale orders, 6 cents.

MAPS SHOWING NATIONAL PARKS AND MONUMENTS

Chaco Canyon, N. Mex. This monument embraces 20,629 acres, set aside in 1907 to preserve the extensive prehistoric communal or pueblo ruins along the canyon of Chaco River in San Juan and McKinley Counties. These ruins are of extraordinary interest because of their great number and size and because they contain innumerable and valuable relics of the prehistoric people who built and lived in them. Chaco Canyon is shown on the Largo map, one of the standard maps described on page 1. Limiting parallels, 36° and 37°. Limiting meridians, 107° and 108°. Scale, 1:250,000, or about 4 miles to 1 inch. Contour interval, 200 feet.

El Morro, N. Mex. This monument, set aside in 1906, has an area of 240 acres and comprises two lofty rocks in Valencia County, known as El Morro and Inscription Rock. The monument is shown on the Wingate map, one of the standard maps described on page 1. Limiting parallels, 35° and 36°. Limiting meridians, 108° and 109°. Scale, 1:250,000, or about 4 miles to 1 inch. Contour interval, 200 feet.

Gila Cliff Dwellings, N. Mex. This monument, set aside in 1907, contains 160 acres and consists of a group of cliff dwellings known as the Gila Hot Springs Cliff Houses, situated on public land in the Mogollon Mountains, within the Gila National Forest, Socorro County. These ruins are of exceptional scientific and educational interest, being the best representatives of the cliff dwellers' remains in the southwestern part of the United States. The monument is shown on the Alum Mountain map, one of the standard maps described on page 1. Limiting parallels, 33° and 33° 30'. Limiting meridians, 108° and 108° 30'. Scale, 1:125,000, or about 2 miles to 1 inch. Contour interval, 100 feet.

Mesa Verde National Park, Colo. This park, comprising 48,966 acres, was set aside by Congress in 1906, in order to preserve the remarkable ruins of prehistoric cliff dwellings discovered in canyons tributary to Mancos River on the Mesa Verde, Montezuma County. The plan and construction of these ancient structures and the stone implements found in them show that their builders had reached a high order of civilization. The three largest ruins have been named Cliff Palace, Spruce Tree House, and Sun Temple. Sun Temple was not discovered until 1915 and is of a type hitherto unknown in the park. It was undoubtedly used for religious ceremonials and is regarded as the highest example of Mesa Verde architecture. Cliff Palace and Spruce Tree House were community dwellings, each providing homes for at least 350 persons, and containing—besides living quarters—mill rooms, storage rooms, granaries, ceremonial rooms, and towers. It is estimated that these structures were erected as early as 1300 A. D.

The Mesa Verde National Park is shown on a special map of that name, listed on page 8. A portion of the park is also shown on the Soda Canyon map, one of the standard maps described on page 1. Limiting parallels, 37° and 37° 15'. Limiting meridians, 108° 15' and 108° 30'. Scale, 1:62,500, or about 1 mile to 1 inch. Contour interval, 50 feet.

Rocky Mountain National Park, Colo. This park, which was established in 1915, is destined to be visited by a great number of people because it is within easy reach of many large cities. The park covers 258,982 acres and contains some of the finest scenery in the Rocky Mountains—snowy peaks above the clouds, steep-walled canyons, flower-bedecked valleys, and mirror-like lakes. Longs Peak, the loftiest mountain in the park, is 14,255 feet above sea level. The park is of great interest to geologists on account of its striking evidences of glacial action in prehistoric time.

The Rocky Mountain National Park is shown on the Rocky Mountain National Park map, one of the standard maps described on page 1. Limiting parallels, 40° and 40° 33' 20''. Limiting meridians, 105° 30' and 106°. Scale, 1:125,000, or about 2 miles to 1 inch. Contour interval, 50 feet.

Wheeler, Colo. This monument, set aside in 1908, consists of certain volcanic formations within the Rio Grande and Cochetopa National Forests, Mineral County, of unusual scientific interest because they admirably illustrate erratic erosion. The monument covers 300 acres and is shown on the Creede map, one of the standard maps described on page 1. Limiting parallels, 37° 30' and 38°. Limiting meridians, 106° 30' and 107°. Scale, 1:125,000, or about 2 miles to 1 inch. Contour interval, 100 feet.

Lincoln Lore

Bulletin of the Louis A. Warren Lincoln Library and Museum. Mark E. Neely, Jr., Editor.
Mary Jane Hubler, Editorial Assistant. Published each month by the
Lincoln National Life Insurance Company, Fort Wayne, Indiana 46801.

August, 1978

Number 1686

Pale-faced People and Their Red Brethren

It was inevitable. The civil rights revolution led to a spate of works on Lincoln and the Negro. When the civil rights movement spilled over into crusades for other kinds of people, Lincoln scholarship could not be far behind. The American Indian movement now has its angry equivalent of Lerone F. Bennett's "Was Abe Lincoln a White Supremacist?" (*Ebony,* XXIII [Feb., 1968]). David A. Nichols's *Lincoln and the Indians: Civil War Policy and Politics* (Columbia: University of Missouri Press, 1978) is less journalistic and more scholarly than Bennett's uncompromising attack on Lincoln, but, fundamentally, it makes the same unreasonable demand that Abraham Lincoln live up to this century's definition of humanitarianism.

The chapter titles constitute the headings of an indictment: "The Indian System: 'A Sink of Iniquity,'" "Lincoln and the Southern Tribes: 'Our Great Father at Washington Has Turned Against Us,'" "Indian Affairs in Minnesota: 'A System of Wholesale Robberies,'" "Lincoln and Removal: 'A Disagreeable Subject,'" "The President and the Reformers: 'This Indian System Shall Be Reformed,'" "The Failure of Reform: 'The Do Nothing Policy Here Is Complete,'" "Concentration and Militarism," and "Lincolnian Attitudes Toward Indians: 'A Dying Race ... Giving Place to Another Race with a Higher Civilization.'" The tone of the book is indignant, and the message, as with almost all modern books on Indian policy in the nineteenth century, is depressing.

What Nichols proves

and what he laments are two different things. The record of the United States government in Indian policy during the Civil War was deplorable as usual. Lincoln's culpability for this record, however, is not so clearly delineated.

No book in the field yields so clear a view of the developments in Indian affairs during the Civil War. There were really several different Indian problems, each of which ran its course to a different unhappy ending. The Southern tribes (or Five Civilized Tribes), resident by the time of the Civil War in Indian Territory (present-day Oklahoma), were peculiar in that they held Negro slaves and were close to the Confederacy geographically. Despite treaty obligations to protect the tribes on their reservations, the United States abandoned the tribes, who made alliances of convenience with the Confederate States of America. Loyal Indians led by Creek Chief Opothleyaholo fled to Kansas, where they lived the miserable life customary for all war refugees.

Late in 1861, the administration decided to retake the reservations, and by January of 1862, it was decided to use Indians as soldiers in the campaign. Nichols notes that this decision did not have the far-reaching effect of leading to citizenship for Indians that the decision to use Negroes as soldiers would have. He does not give a full analysis of the reasons for the difference in result, but speculation on the subject is illuminating. In the first place, Indians were not vitally and logically linked to the Civil War, as Negroes were. The Indians

From the Louis A. Warren
Lincoln Library and Museum

FIGURE 1. Creek Chief Opothleyaholo in a youthful portrait painted long before he led loyal Indians to Kansas in the Civil War.

veterans. Third, the were not e〼 of the to very
a〼t and it wa wi〼 a〼 tha the nhe wee
diminishing towards e〼 The was little ne〼 to be
c〼 about the future of the I〼a in Ame〼 so〼y ;
he h〼 no future.

The Battle of R〼 Ridge, Arkansas, in wh〼 a n〼r of
I〼〼 fought for the Confederacy, was a defeat for the Cn-
federacy wh〼 c〼e an a〼 1 of I〼a territory.
T〼 l〼y refugee problem was no s〼v t〼e
government had to p〼 to se〼 th〼 bi〼 and p〼 to protect
t〼m o〼 they wee t〼e. In 1864, the government re〼le
t〼e refugees fr〼 Kansas, too l〼e for planting seas

A separate I〼〼 problem wa the Sioux uprising in
M〼 in 1862 N〼〼 d〼y ab on third of his
bi〼 to this famous e〼 in L〼n I〼〼 re〼 virtue of his a〼 lies n〼 o〼y , its thorough grounding in
m〼〼 sources but a〼 in its treatment of the Si〼 up-
rising, n〼 as an il〼 a〼 s〼〼 event, but as a
part of the Lincoln administration's continuing development.
Nichols's account is particularly useful in showing the reso-
lution of Indian pr〼l after the fam〼 hangings in Man-
kato, Mi〼〼 te the day after Christmas, 1862 (see *Lincoln
d, re* N〼e 1627 and 1628). The wa〼 interested Lincoln for
the first t〼n se〼y in Indian reform, but the re〼 of
the Minnesota problem involved no reforms. M〼〼 offi-
cials and the national government assuaged local re〼e
mn〼 over L〼〼 pardoning 265 Sioux prisoners by
removing the tribe fr〼 the state and keeping the pardoned
Indians in co〼e 1 1 The governmentals removed the
Winnebagos, wh had not participated in the uprising, but let
the Chippewas stay, probably because they were of sp〼l interest to Indian reformer Henry B. Whipple, who had in-
flue〼 wi〼 the L〼n administration.

By 1864, L〼n had lost interest in Indian ref〼 rm. The wa〼
and reelection preoccupied him, the Indian C〼i〼 Dole
tried a policy of concentrating the Indiansо n a few rese〼 -
tion re〼n fr〼 wh〼 settlement, and th〼 n〼y played a
larger role th〼n before in dealing w〼h Indians. The Army
proved as inept at handling Indians as the Interior D〼p -
m〼n notoriously corrupt Off〼 of Indian Affairs. In
November, 1864, at Sand Creek, Colorado Territory, wh〼
militia massacred hundreds of Indians, killing children,
scalping w〼m c〼t〼 1 men, and butchering pr〼〼

In most instances, because of Lincoln's inattention to
Indian affairs, Congress played a major role in Indian policy.
The settlement of Minnesota's Indian problems, which
Nichols characterizes as "Trading Lives for Land and
Money," was embodied in legislation passed by the United
States Congress. Congress gave Minnesota a $1.5 million in-
demnity for losses incurred in the war. Congress appro-
priated the money to remove the Sioux from Minnesota. Con-
gress appropriated money to remove the Winnebagos from
Minnesota. If this was a "Lincoln bargain," as Nichols de-
scribes it, it was a bargain on which there was widespread
agreement in Washington, D.C.

Often, Nichols assumes that Indian Commissioner William
P. Dole's policies were Lincoln's policies. Were Salmon P.
Chase's Treasury Department appointees who opposed Lincoln's renomination in 1864, Lincoln's appointees? One must
be careful in judging the "Lincoln administration" or "the
government." In fact, it remains difficult to describe Lincoln's Indian policy because he made so few statements on the
problem and because he took little direct action in Indian
affairs.

Barnum's American Museum.
Sioux and Winnebago

Indians' problems.

The book's one-sidedness can best be seen in its treatment of the formulaic language of Indian relations. This mannered, formal pidgin-English seems quaint and has always troubled historians of Indian relations. In the hands of a historian with a case to make, it can be a powerful tool. Nichols, probably ... us ... by ... his ... to interpret it seriously when us ... by Indians. ... faced ... and their red brethren," when a ... of ... isited the White ... on Mch 27, 1863, is ... an "incredible ..." by N ... ls. By contrast, Nichols says this of a Cherokee ... of fealty:

In spite of his ... of ... cause, the ... died to place faith in the White House after ... Johns ... ame fine. "Our trust is in your ... and sense of j ... to ... us from wrong and ... That trust in the "great father" was ... to be ... more ... for the ... far-ther ... with in the Republican state of ...

... is no more ... to ... formal ... ... than there is to ... usually ... expressions of ... of brotherhood ... red men and ... is a ... they, in t day's ... of ... by for the Indians to treat ... side of the story with the historian's sual critical tols.

The angry tone and ... for high effect by ... the ... with ... in ..., ... this ... It is otherwise a well-re-... my ... less of the major develop-... in ... the Indian. ... of ... Press, deserves special praise for a beautifully designed and carefully ... He typeface is ... few ... at the ... of the ... is ... the ... ... and the jacket ... have ... in the country. Nichols's *Lincoln ad the Indians* lifts a ... in the Indian ... will ... (at ... with ... author's ... to ... an ... Lincoln's

Nichols's brief treatment of Lincoln's personal experience with Indian affairs before entering the White House typifies his grudging interpretation of Lincoln's actions. He mentions the famous ... in the Black Hawk War in ... coln allegedly defended an old Indian who ... into ... from ... to kill ..., but he ... the story on ... Sandburg's *Abraham Lincoln: The Prairie Years.* Ben-... ran P. ... has ... more reliable ... for the story. In *Abraham Lin: A Biography*, ... coln let the story stand in a ... with he carefully corrected for William Dean Howells. Nichols concludes that ... learned how to use Indian affairs for po-... that experience is not easily interpreted. In fact, ... turned from the war so late in the ... that he had only two weeks to cam- ... for the legislature. Moreover, ... enlisted, in ... for the ... (or but to be) and ... no family in New Salem. He may ... tacy of the Indian- ... image in the age of Andrew ... but ... ever tried to ... on such an ... He did not go by the ... my ex-frontier ... did ..., ... had he confessed plainly that he never saw ... fighting Indians" in the war. ... he also prided himself on his ... praise ... seems in ... on his experience, ... as it no way to ... as an Indian-fighter. ... in the ... he ... the president," Nichols says, "apparently never challenged the American co... on the ... for ... to ... way for ... gress." This is ... Nichols's ... to challenge the ... a ... of ... His ... His ... the Indian ... suffer; ... that the Indian ... Everything in Lincoln's government ... on the patronage ... Bureau was a ... of the ... Indian affairs," was an ..., even the war. To ...

CPSIA information can be obtained
at www.ICGtesting.com
Printed in the USA
BVHW041025210219
540828BV00009B/136/P

9 780332 704197